THE BEAUTY OF STRAW BALE HOMES

THE REAL GOODS SOLAR LIVING BOOKS

The Beauty of Straw Bale Homes by Athena and Bill Steen

Serious Straw Bale: A Home Construction Guide for All Climates by Paul Lacinski and Michel Bergeron

The Natural House: A Complete Guide to Healthy, Energy-Efficient, Environmental Homes by Daniel D. Chiras

The New Independent Home: People and Houses that Harvest the Sun, Wind, and Water by Michael Potts

Wind Energy Basics and *Wind Power for Home & Business* by Paul Gipe

The Earth-Sheltered House: An Architect's Sketchbook by Malcolm Wells

Mortgage-Free! Radical Strategies for Home Ownership by Rob Roy

A Place in the Sun: The Evolution of the Real Goods Solar Living Center by John Schaeffer and the Collaborative Design/Construction Team

The Passive Solar House: Using Solar Design to Heat and Cool Your Home by James Kachadorian

Independent Builder: Designing & Building a House Your Own Way by Sam Clark

The Rammed Earth House by David Easton

The Straw Bale House by Athena Swentzell Steen, Bill Steen, and David Bainbridge with David Eisenberg

Real Goods Solar Living Sourcebook: The Complete Guide to Renewable Energy Technologies and Sustainable Living, 10th Edition, edited by Doug Pratt and John Schaeffer

Real Goods Trading Company in Ukiah, California, was founded in 1978 to make available new tools to help people live self-sufficiently and sustainably. Through seasonal catalogs, a periodical (*The Real Goods News*), a bi-annual *Solar Living Sourcebook,* as well as retail outlets and a Web site (www.realgoods.com), Real Goods provides a broad range of tools for independent living.

"Knowledge is our most important product" is the Real Goods motto. To further its mission, Real Goods has joined with Chelsea Green Publishing Company to co-create and co-publish the Real Goods Solar Living Book series. The titles in this series are written by pioneering individuals who have firsthand experience in using innovative technology to live lightly on the planet. Chelsea Green books are both practical and inspirational, and they enlarge our view of what is possible as we enter the new millennium.

Stephen Morris
President, Chelsea Green

John Schaeffer
President, Real Goods

THE BEAUTY OF STRAW BALE HOMES

Athena and Bill Steen

White River Junction, Vermont

The photograph on page 77 is by Miquel Fairbanks.

Designed by Ann Aspell.

Library of Congress Cataloging-in-Publication Data

Steen, Athena Swentzell, 1961–
The beauty of straw bale homes / Athena and Bill Steen.
p. cm.
ISBN 1-890132-77-2 (alk. paper)
1. Straw bale houses. 2. Building—Details. 3. Straw bale houses—Pictorial works. 4. Architecture, Domestic.
I. Steen, Bill. II. Title.

TH4818.S77 S74 2000
693'.997—dc21 00—59648

Chelsea Green Publishing Company
Post Office Box 428
White River Junction, VT 05001
(800) 639-4099
www.chelseagreen.com

Printed in China on chlorine-free European matte paper.

03 02 2 3 4 5

CONTENTS

PREFACE

OUR FIRST BOOK, *The Straw Bale House,* was very much an overview of the methods being used by straw bale builders at that time, the mid-1990s. The present book, although mostly made up of color photographs, is similarly a sampling of the many different types of straw bale buildings that have been built. This book is dedicated to the ingenuity and vision of all those who had the courage to try something different, and thereby contributed to the evolution of an imaginative building technique.

The task of selecting a group of buildings to represent a variety of climates and styles was challenging, almost impossible. Our travel is limited in that we have small children and prefer to stay close to home. Consequently most of our examples were either places we could easily reach or ones for which friends and builders could provide good photos. Ultimately, we settled on a selection from North America, and decided to save the many straw bale buildings that are appearing in faraway places such as Europe, Russia, Australia, China, and Mongolia for a future volume focusing on straw bale buildings around the world.

We also included a number of brief, explanatory essays, which contain information we felt important for the good performance and durability of straw bale buildings. These are not technical guides by any means, but are intended to point the reader in the right direction. For fun we have also included several simple recipes for natural paints and plasters.

Acknowledgments get more and more difficult as time goes by. We continue

learning from just about every person who becomes involved in this field, whether they are newcomers or experienced builders, architects, and engineers. It is the sum total of their combined efforts, successes and failures, and beautiful structures that have made this book possible. Our thanks to everyone.

Athena and Bill Steen

DESIGNING A BEAUTIFUL HOME

WHAT MAKES a house beautiful? Is it complexity of design, square footage, or money spent? Is it the prominence of a "view-obsessed" starter castle on top of the hill? Or the prestige of having an architect's "showcase"? Perhaps beauty originates somewhere beyond all that, in elements more internal and more related to the human heart. Isn't beauty closer to the fullness you feel when sitting in a cozy corner? Or the feeling you have after working hard all day, then standing back to look at your work with a childlike pride? Beauty is more easily found in that which has been touched with love and care than in polished details and lavish furnishings.

For us, the most memorable houses have been small and simple, and those with which the owners have played an integral part. Many of these are not finely finished, nor are they examples of superb craftsmanship, but they are rich with character and ingenuity. All too often, homeowners relinquish to the architect and builder the opportunity to be personally involved in the creation of their home. From a distance, they monitor the building and the shaping of spaces they will one day inhabit. Someone else designs it, another person builds it, and in the end they are left with a house for which they did little else than pay. Most modern houses are not much more than purchased commodities that tell of our increasing dependence on specialists and our loss of ability to craft our own lives. Most of such houses are impersonal and empty of feeling. This is not to say that architects and builders don't have a place in the process of making homes, but simply that there is a tendency to over-rely on them.

Speed and efficiency dominate conventional building. The builder is expected to build as fast as possible for the least amount of money. Under those stressed circumstances, there is very little time or space for intimate creativity. To stay on schedule and within budget, it is presumed necessary to rely primarily on mass-produced industrial materials that don't encourage deviation from their intended use and that require the extensive use of power tools. The best-intentioned designers and builders find themselves confined to a narrow range of possibilities, and not surprisingly the results are often boring and predictable.

On the other hand, we have discovered in our own work that using natural and local materials encourages a very different way of building. These materials often have an inherent beauty that stands out without the need for complex forms and shapes. Lacking uniformity, rigidity, and angularity, they naturally lend themselves to soft, organic curves. They resonate with

the textures and colors of their natural context, reminding us that the building belongs where it is placed. They are forgiving, and beg to be sculpted by hand or worked with simple tools, opening the door for more people to be involved in the process.

Although many houses built from natural materials are beautiful, the use of such materials does not guarantee that a house will be beautiful. We have seen many straw bale houses that feel no different than any other building. When incorporated into conventional construction, natural materials become subjugated to the same stresses and patterns. Straw bale walls can rapidly become a very insignificant part of the whole house. We once asked a friend how she liked her new house, and she responded, "I really wanted a straw bale home, but what I got was a house with straw bales in the walls." Her house was beautiful in the typical meaning of the word. It was well built and nicely finished. And yet it lacked the texture of her presence. Its abstract perfection kept everyone at a distance, including her.

The walls surrounding us day in and day out need to embrace us, our dreams and passions woven into their very fabric. They need to sing the story of who we are. Otherwise our houses will never become our homes. And it is in the depths of this magical transformation where genuine beauty lies.

THE BEAUTY OF STRAW BALE HOMES

STRAW AND HAY: THE TEST OF TIME

It wasn't long after the appearance of baling machines in the 1850s that straw and hay bales began to be considered a building material. Historical patents for bale walls date back to the 1880s in the state of Indiana. The first significant use of bales as a building material occurred in the Sandhills of Nebraska, a vast tract of desolate, grass-covered dunes. An abundance of wild grasses and the lack of timber and good building soils provided incentives to devise new building techniques using unconventional materials. The oldest bale building on record is a school built in Scott's Bluff County in 1886 or '87, which cattle ultimately ate.

We have since found examples of buildings scattered throughout the United States, Canada, and as far away as France, with rumors of others around the world. We thought it appropriate to begin this book with a few examples of older buildings that remain in good condition and are still in use today.

FAWN LAKE RANCH

THIS RANCH near Hyannis, Nebraska, has two of the oldest hay bale buildings in existence, which were built somewhere between the turn of the century and 1914. The main house and the bunkhouse were built of baled meadow hay using a stationary baler, and are still in use today. In the 1950s, when a wall was opened up to make way for an addition, some of the old bales were set to the side near a corral where horses on the other side reached over the fence and began eating the hay, which was about fifty years old. When we paid a visit in 1997, we were able to remove a panel of masonite covering one of the walls to reveal the bales behind it, still in good condition. The interior walls have paneling that appears to have been mounted on 2-by-4s that were attached to the bale walls with wire.

Left: Possibly the oldest bale building still in use today.

Right: Wall covering removed, showing the good condition of the original bales.

THE DIFFERENT TYPES OF STRAW BALE WALLS

Builders and designers have traditionally divided straw bale wall-building methods into two basic categories, *load-bearing* and *infill*.

Load-bearing has come to imply that the bale walls, without the aid of a structural framework, will carry:

– vertical or gravity loads (the roof, upper floors, snow, and sometimes water).
– lateral loads (wind).
– seismic loads (every which way).

Infill implies the use of a structural framework in combination with straw bale walls. In contrast to load-bearing walls, with infill designs:

– vertical or gravity loads are carried by the structural system, such as wood framing.
– lateral and seismic loads are partially carried by the straw bales, and also transferred to that main structural system.

In all cases the bales support their own weight and often the weight of windows and shelving. Straw provides the primary layer of insulation and serves as the substrate for the interior and exterior wall finishes.

Is it necessary to choose one method or the other? Which of the two is better? It depends on the size and complexity of the building, the loads to be countered, the type of wall finish, and the local climate. We have found that integrating features of both load-bearing and infill into the walls of the same building has many advantages and often creates a better wall than either method used separately.

We have only begun to explore how to build walls out of this unique material, and with additional experience and experimentation, it is clear that the future will yield more creative and efficient wall systems.

THE MARTIN/ MONHART HOUSE

Built in 1925 in Arthur, Nebraska, this simple building exemplifies the classic "Nebraska Load-Bearing Style." The small size of the house combined with the modest number of window and door openings make its design ideal for load-bearing walls. The hip roof distributes weight around the walls of the building and keeps a low profile. An interesting note is that the house was covered with earthen plaster until receiving a coat of cement stucco in 1930. It remains a family museum and is open to the public

This house is typical of many early Nebraska-style buildings.

CHUCK AND MARY BRUNER'S HOUSE

Chuck Bruner in front of his home. "I built it out of straw because I had been inspired by the straw bale grocery store nearby in Glendo, and since we really didn't have much money, it was the cheapest way possible to build a house," said Chuck. The house has never shown major cracks despite a frost level typically five feet deep, and it withstood a 5.5-level earthquake without damage.

Chuck and Mary Bruner built their 1,300-square-foot straw bale home in 1949 in Douglas, Wyoming, where they still live today. "We hand-loaded all the sand and gravel and mixed all the concrete and mortar with a mixer that we had rigged with a washing-machine engine." All the straw was baled with a stationary baler, then hand-tied. The house has a post-and-beam structure that uses 6-by-6 posts with mortared bales, and took about five months to build while Chuck was working nine-hour days, six days a week as an auto mechanic. He and his father did the majority of the construction. The exterior has been maintained with limewash prepared from carbide lime.

ANDRÉ DE BOUTER

THE FEUILLETTE HOUSE

WHEN BALES FIRST appeared, they evidently appealed to more people as a potential building block than we realized. This two-story straw bale house in Montargis, France, was designed and built in 1921 by a Mr. Feuillette, who intended it to be a low-cost prototype for reconstruction of farms and peasant housing after the First World War. The house uses bale infill with a timber-frame structure. The exterior finish is a cement, lime, and sand render that has been covered with Virginia creeper for at least twenty years. A bitumen coating was applied to the top of the foundation. The roof, covered with terra-cotta tiles, is insulated with wood chips.

The oldest known straw bale building in Europe. The Virginia creeper gives additional protection to the stucco-covered walls.

TRADITIONAL MEXICAN STRAW HOMES

LONG BEFORE BALERS, straw was used for building—mostly to thatch roofs, but also to construct walls. While helping conduct a workshop in the state of Tlaxcala, Mexico, we found these unique houses in a town where this style of building has been practiced for as long as anyone could remember. The owner-builder, Don Alejandro, told us that his houses are about twenty-five years old. Looking at these buildings, one can't help but think about the ability of straw to withstand years of weather without any protective covering or coating. In contrast, the wrong choice of modern wall-surfacing materials and coatings can cause rapid deterioration of the straw if moisture becomes trapped within a wall cavity.

Above left: Straw houses, Tlaxcala, Mexico. The roof structures were made using stalks of the century plant (agave). The straw is rye, or *centeno* as it's called locally.

Above right: Don Alejandro completing the ridge detailing.

Figure of heart woven from stalks of wheat straw.

THE RIGHT STRAW

Bales used for building should be dense and compact, capable of supporting a substantial amount of weight without changing shape or deforming. The strings should be tight, holding the bale securely together. Most importantly, straw used for building should be bright, golden yellow with no signs of discoloration, which indicates moisture damage; they should have been stored under cover for protection from rain, because once moisture has penetrated to the interior of a bale, it is almost impossible for drying to occur without some deterioration. Simply put, use only bales that have always been dry and kept out of the weather. Try to avoid justifying to yourself that it's okay to use rain-damaged bales for building.

PAM TILLMAN'S HOUSE

A PROJECT OF Design and Building Consultants, this house was built for Pam Tillman as a rental unit in the historic *barrio* section of Tucson. At 900 square feet, it is a small and beautiful L-shaped building. Featured in *The Straw Bale House,* this was one of the first straw bale buildings to challenge the assumption that load-bearing straw bale walls used less wood for construction. It was originally designed as a load-bearing building, but when comparisons were made, it was determined that it would be more efficient to use a "modified post-and-beam" system by developing the door and window bucks as integral members of the bearing structure.

Right: John Woodin, who has become one of the more accomplished straw bale builders in the Tucson area, was introduced to straw while working for Pam on this project. Not only did working with straw appeal to him, he and Pam have been partners ever since.

Left: Porch, seen from the front.

JULIE HARDING'S HOUSE

JULIE HARDING had Tucson builder John Woodin build her straw bale house, which was designed by architect Paul Weiner to fit into a historic neighborhood where most of the buildings have a similar "Tucson Victorian" or "Desert Bungalow" style. Different than most straw bale houses in the area, Julie's home has wood floors and wood-framed door openings instead of beveled plaster. It is 2,500 square feet, with two bedrooms and baths.

Facing page: The exterior is finished with stucco and the interior with a smooth gypsum plaster.

Right: The yellow color heightens the light-gathering qualities of this spacious kitchen.

Above left: Wood-framed doorway, unusual in a straw bale house.

Above right: Living room. All the colorful fireplace tiles were hand-made by John Woodin.

AN ARIZONA RANCH CHAPEL

THIS BUILDING exemplifies how much fun can be had building with straw. Many of the visually pleasing features of this fabulous building were devised by accentuating natural characteristics of the bales themselves. While we have come to really appreciate the chapel's singularity and artistry, it also represents the paradox in early bale-building design. Much of the appeal of straw bale walls has come from their resemblance to other forms of thick-walled masonry, such as adobe, so there have been numerous attempts to use bales to build imitation adobe structures with parapet walls, which extend up above a flat or pitched roof and are typically covered with stucco, brick, or metal flashing. This design is not appropriate for straw bale buildings, as the lack of overhangs leaves the bales unprotected and vulnerable to weather, and places a lot of unnecessary stress on the walls. As straw bale construction continues to evolve, the most successful buildings will be those designed to utilize the unique capabilities of straw, functionally and aesthetically.

Despite their traditional look, parapet roofs are too susceptible to moisture to be used with straw walls.

The interior forms in the chapel were created with the moldable and irregular shapes of the straw bales. The ceiling is made of ocotillo cactus. Patsy Lowry designed and painted the decorative artwork here and on the exterior (opposite).

JOHN POSELEY AND MARGARET O'DONNELL'S HOUSE

Located in the middle of the Arizona desert near Phoenix, where summer temperatures frequently exceed 110 degrees Fahrenheit, John and Margaret's home requires a minimum of air conditioning to remain comfortable. Their neighbors, living in conventional homes of similar size, commonly have summer electric bills that exceed $400. In contrast, John and Margaret's highest electric bill has been only $60, during their first summer, proving how effectively straw bales insulate a house from heat as well as cold.

Facing page: One of the most interesting aspects of this house is John's design: The house has a wraparound porch that also functions as the building's load-bearing structure, with roof trusses spanning from porch to porch. This eliminates the need to integrate bale walls into the structural framework. The porch provides protection from rain and moisture all the way around the building, and shades the ground to reduce the heat at the building's perimeter. Placing the structure outside the walls is not recommended in more severe climates, where this exposed structure would deteriorate more rapidly than if incorporated protectively in the walls.

THE R-VALUE OF STRAW BALES

R-value means "resistance value," using a rating system for measuring the relative capacity of insulating materials to resist heat transfer.

The first R-value tests of straw bales were conducted on individual bales by Joe McCabe; he achieved results that ranged on average, depending on the size and orientation of the bale, from R-40+ to R-50+. Later tests conducted on whole wall assemblies of straw bales got lower results: Oak Ridge National Laboratories had R-31.2, while the California Energy Commission adopted a value of R-30 based on testing conducted by Nehemiah Stone. These results were disappointing to some, who thought that the tests for wall assemblies, including allowances for losses, would be similar to results for individual bales. However the same disparity holds true for other materials. For example, 2-by-6 framed walls that are nominally rated at R-19 only achieved whole-wall ratings of R-12.8 at Oak Ridge even with properly installed insulation (which it often is not). By these terms, straw bale walls compare very impressively.

It's also very important to remember that laboratory tests are conducted under steady-state conditions. In the real world of nature, consistent conditions rarely exist for more than a minute or two. The true test of any material is what happens when one lives in a house and keeps track of the energy required to maintain a reasonable level of comfort. Rather than measuring just R-values, it is essential to consider the overall contributions of a bale wall's thickness and the thermal mass provided by interior and exterior coatings. Together these provide comfort and thermal performance much greater than R-value alone would indicate.

THE CANELO PROJECT: ATHENA AND BILL STEEN'S HOME

Fresco color on lime-plastered wall.

VISITORS TO OUR home often assume that we live in a straw bale house, and are instead surprised to learn that the thick walls of our house are made from adobe. Yet it doesn't take long for guests to appreciate how extensively a few simple materials, including clay plasters, casein paints, cob, and bales, can transform an old building into something with character and beauty. Since the turn of the century, several different families have invested much of their lives and energy here, and we have come to accept that we are part of the same process. What we once saw as a chaotic arrangement of space has become a rich opportunity to create a very comfortable and unique living environment.

Because we regularly host straw bale workshops and other training sessions through our nonprofit organization, the Canelo Project—covering such topics as natural plasters, earthen floors, clay ovens, and paints and washes—we have also constructed a number of small cottages and demonstration structures. These have afforded us additional opportunities to experiment with an ever-changing palette of materials and techniques.

We prefer the small, the simple, and the handcrafted. Much of the work we do involves the use of clay for plasters, floors, paints, or bricks. Not only is clay accessible to people all over the world at little or no cost, but its sculptural qualities engender a different and more artistic way of building.

The Steen family: Bill, Benito, Athena, Oso, Arin, and Kalin, sitting on porch benches that are plastered bales finished with a casein paint.

Left: Cottage interior with adobe partition wall and seat, finished with a combination of casein and clay paints.

Below: Kitchen corner with cob fireplace and benches. The most used corner of our house: at times we pack up to fifteen people around the table.

Above and right: A playful structure designed to encourage students to experiment and practice on smaller projects, it also demonstrates the ease and fun of sculpting with clay in combination with bale walls. The seat was built using hand-formed cob bricks with packed stone and rubble-fill underneath. The roof has rafters of local agave or century plant, with purlins of bamboo covered with a layer of greenhouse plastic and water-reed mats. The walls are finished with clay and lime plasters as well as casein and clay paints. The west wall was left uncovered to reveal the straw bales and exterior bamboo pins, like a giant "truth window." The yellow wall is lime plastered, frescoed with an ochre mineral pigment.

MATTS MYHRMAN AND JUDY KNOX'S RETROFIT

Above: Three-string straw bales were placed on edge and anchored to the existing masonry wall with wire ties. An earthen plaster with a high proportion of straw was used to cover the bales around the building. Lime plaster was used over the earthen plaster anywhere the walls were not protected by porches, and also used for detail work around windows and doors.

Facing page: The improvement in comfort and overall appearance is profound.

MODERN STRAW BALE pioneers Matts and Judy live on a small urban lot, where they originally built the first permitted load-bearing bale building in the city of Tucson. They recently directed their efforts to an older concrete block home that they owned on an adjacent lot, a house typical of many built in the 1950s. The building had little to no insulation and was very uncomfortable during much of the year, despite the mild Arizona winter. Considering their own involvement in the straw bale movement, they felt embarrassed to be owners of such an "energy hog," and decided to embark on a thorough renovation.

Their strategy consisted of wrapping the walls with straw bales, increasing the roof insulation, installing energy-efficient windows, adding porches on two sides, and plastering the entire building.

RETROFITTING EXISTING BUILDINGS

Many people dream of building an incredible new house out of sustainable materials without realizing that many of their wishes could come true through renovating the very place where they presently live. Until now, builders and designers have given little attention to the possibility of retrofitting existing buildings with straw bales to improve their thermal performance. Many older buildings were very well built and incorporated much better materials than are typically found in today's buildings. Wrapping straw bales around a poorly insulated building can turn a thermal nightmare into a dwelling of greatly increased comfort and aesthetic appeal, often at much lower cost than building a new one. This can even be done in many urban locations that are very appealing in terms of neighborhoods, convenient access to services, and less time spent in a car.

The procedures for placing straw bales over the existing exterior walls of older buildings have not yet been clearly defined. The two major challenges are attaching the bales to existing walls, and making sure that moisture will not accumulate between the old and new walls. The details will vary somewhat depending upon climate and the type of wall and finish the bales are being placed against. The direction of vapor flow will need to be considered, so that the bales are not placed against impermeable walls where moisture vapor will be trapped.

To provide protection against fire as well as insects and rodents, bales should normally be plastered on both sides, even when they are subsequently covered with siding or rain screens. One way to provide this protection and also attachment to the existing wall is to stack the bales up against a fresh coat of plaster. Another option is to coat the sides of the bales that will be set against the existing wall and allow them to dry before placement, relying then upon some other method of attachment.

Above: Before and after the retrofit.

Matts Myhrman.

BOB COOK AND FRIEDERIKE ALMSTEDT'S GUEST HOUSE RETROFIT

Above: Bob and Friederike beside their homemade, earthen-plastered fireplace.

Bob and Friederike bought a fifty-year-old concrete-block home conveniently located near Tucson city center, close to Bob's favorite lumberyard. Unattractive, but built of solid materials, the house was inexpensive and offered them a more sustainable and affordable option than acquiring an undeveloped piece of property and building a new home. Location was important, as they rely on biking and walking for much of their in-city transportation. Their first project was building a 300-square-foot straw bale workshop in their backyard, which was featured in our book *The Straw Bale House.* To help block out noise from a major street in front of the house, they added a straw bale privacy wall around their property that also created numerous outdoor sitting areas along the perimeter.

Their priority in retrofitting the main building was super-insulating the walls and roof to protect them from the summer heat of the Sonoran Desert. They also installed energy-efficient windows, constructed a south-facing solarium for solar gain, and converted a covered porch called an "Arizona room" along the west wall.

Three-string bales were laid on edge on a new concrete foundation and joined to the existing masonry walls with straw-clay mortar, which in the hot, dry climate of Tucson could dry without creating moisture problems behind the bales. Cement stucco was chosen as the exterior finish and coated with an elastomeric paint to prevent penetration of moisture into the interior of the wall where it might create problems between the bales and the latex-painted concrete block.

Above: Cook house before the retrofit.

Right top: Privacy wall. Through every stage of the work, many recycled and discarded materials were used, especially lumber. Gutters and cisterns were also installed, for water harvesting from all roof surfaces.

Right: The remodeled interior, painted with homemade casein paint.

RECIPE FOR CASEIN PAINTS

"I feel empowered," Bob Cook jokingly told me over the phone as he was painting the interior of their retrofitted home with casein paints that he and Friederike were mixing at home. "Here I am, mixing as much paint as I need, and it is costing me next to nothing. It's an amazing feeling to know that I don't have to go to the store to buy my paint."

At first, the prospect of mixing one's own paint can look intimidating, but usually it takes little more than getting started for that feeling to disappear. Any homemade effort often involves a glitch or two the first time, as there is no exact formula for how much pigment to use or how much filler is needed. Yet it will probably take you very little time to come up with a mix that works satisfactorily.

Casein paints lack the plastic surface of modern paints, and especially when using natural pigments, they reflect light in a very different way. The following recipe is for a solid and flat paint. Thinning it slightly can make a nice wash, and thinned even more makes a solution to wipe over wooden surfaces.

Facing page: The translucent casein wash surrounding this truth window brings out depth and rich color.

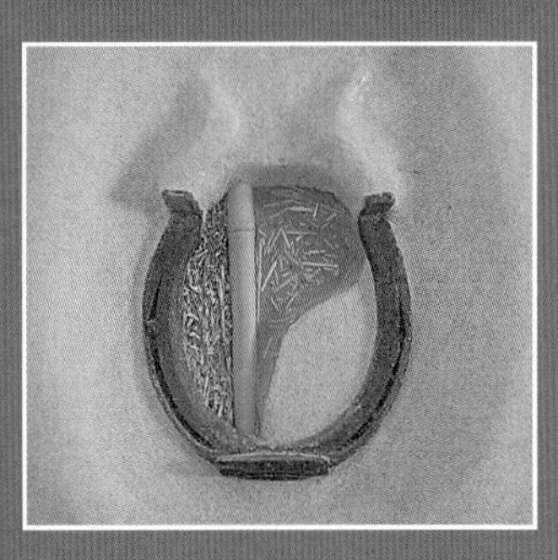

Mix 2 parts casein powder with 8 parts water, then let sit overnight or a minimum of a few hours.

Mix 1 part borax with 2 parts warm water. Combine the borax and casein mix. Add another 6 parts water.

For every 1 part of the above solution, add 9 parts of a solid filler (chalk, powdered marble, white clay, etc.). This will give a flat white paint.

If seeking a color, add pigment to a small amount of the casein solution and mix or grind well. Add this paste or concentrate to the casein and filler combination. Colored clay can be used as a filler and then no pigments will need to be added.

Apply multiple thin coats.

Experiment with thinning the above formula with water. If the mixture dusts, there is not enough casein, and if it cracks and peels, the mixture needs to be diluted with more water.

Casein paint is for interior use, and can be described as water-resistant but not waterproof.

FRANK MEYER

FRANK MEYER

Above: The straw bale addition that serves as the Meyer living room. The front porch is built from cedar gathered in the Austin area. Connecting the straw bale addition to the older building did not pose any particular problems; the main challenge was making the two blend together visually, which they accomplished with a unifying application of cement stucco.

Facing page: Connecting stairway to the original building, built out of Plyboo, a laminated bamboo product, with larger culns used for the railing.

AMY AND FRANK MEYER'S STRAW BALE ADDITION

HAVING DISCOVERED how much time they were spending in their car, Amy and Frank Meyer took a chance and bought a 1950s wood-frame house near the center of Austin, Texas. At the time their new purchase looked like an asphalt-sided abomination. They decided to remodel the existing building and add a 500-square-foot straw bale addition to give them extra room. Frank, owner of Thangmaker Construction, which specializes in straw bale building, built the salvaged-wood, post-and-beam structure with the help of two carpenters, as well as using a bale raising, plastering parties, and friends to complete the rest of the work. They took three years to complete the project, but their relaxed, organic approach gave them a home uniquely suited to their needs.

THE SIVANANDA YOGA CENTER

AFTER A FIRE burned down the headquarters of the International Sivananda Yoga Vedanta Center in the Laurentian Mountains of Quebec, Sivananda decided to replace the original building with a straw bale lodge designed by Michel Bergeron, a Montreal-based designer who co-founded the pioneering design-build firm, ArchiBio. The new lodge is 12,000 square feet with twenty-two rooms, and met the requirements of the National Building Code of Canada. The load-bearing structure is a wooden post-and-beam frame, with bales used as infill insulation, finished with lime cement inside and out. The ground floors have straw-bale–insulated slabs; bales are also used in the roof for insulation, as well as on top as part of a living roof. Almost 10,000 bales were used in the construction of the building. The cost in 1997 was approximately $90 Canadian per square foot.

Designer Michel Bergeron writes: "To me, appropriate design means the creation of a living environment capable of fulfilling all the mental and spiritual needs of the occupants. For this project I wanted to create an atmosphere supportive of the yogic lifestyle."

S. THIRKETTLE INSET: MICHEL BERGERON

MICHEL BERGERON

THE BRISSETTE/PANFILI HOME

BUILT IN 1990, this two-story house in Ile Saint Ignace has approximately 3,000 square feet of floor area. Designed and built by Michel Bergeron, it has a straw bale–insulated slab with a timber-frame structure. Some parts of the load-bearing structure are concealed within the bale wall, while other members of the beautiful red-cedar and hemlock frame are exposed. The bales were covered with a jute fabric pinned with handmade, U-shaped staples. A lime-cement plaster was applied over the jute with a finish coat of lime, white Portland cement, and marble dust, which is refreshed with a lime wash every three or four years. The second floor is stud-frame, insulated with rockwool, and the roof is covered with hand-split cedar shakes. The main source of heat is a Scandinavian masonry heater that doubles as a bread oven and cookstove.

Facing page: This timber-frame structure uses straw bale infill. Exterior walls are plastered with lime and a small quantity of cement.

THE WARBURTON HOUSE

THIS 2,800-SQUARE-FOOT, two-story post-and-beam home has three bedrooms and an open plan organized around a central masonry heater. Straw bales serve as infill insulation. Completed in 1997, it is situated on a heavily wooded 3.5-acre lot bordering the Nottawasaga River in Orangeville, an area called the snow belt of Ontario. The design follows many of the "patterns" outlined in *A Pattern Language,* a marvelous book by Christopher Alexander et al., for instance utilizing sleeping niches, room layouts that follow the movement of daylight, and porches that bridge between outdoor and interior spaces. The posts and beams and much of the flooring, siding, and joists were culled from trees on the property. Willow twigs are used as a decorative siding on the bay windows, and the pigment in the final exterior stucco coating is iron oxide. The house was designed by architect Linda Chapman and built by the owners, who worked as their own general contractors with the assistance of Colin Jones.

Facing page: The dormer windows are detailed with willow-twig siding cut on site. Clear sealer was applied to keep the bark from peeling.

LINDA CHAPMAN

BOB BOURDON

AUBERGE À LA CROISÉE DES CHEMINS

BORDERING THE Rogue River in the pine forest of La Conception, Quebec, the wooden post-and-beam structure of this nine bedroom, straw bale bed and breakfast came from trees milled on site to make room for the building. It was built by Johanne and Bob Bourdon with some architectural design assistance from Montreal architect Maryse Leduc Cummings. Unlike many other straw bale buildings, this one utilizes rough-cut siding with an air space behind it, instead of a plaster or stucco finish.

FINISHES FOR STRAW BALE WALLS

The finish is an integral part of the straw bale wall system, and fulfills a variety of functions. The best finish is a plaster or stucco that bonds directly with the straw. Bonded finishes add strength and support to the wall, control air leakage, provide superior fire control, and help control insects and rodents. When matched to the local climate, they can provide protection from weather and allow diffusion of vapor. Rain screens, siding, and drywall can also be used, but should be applied over straw bales that have been coated with a fully bonded material.

Compressive strength is important in a finish for load-bearing walls, which can continue to settle over time. In general, interior and exterior finishes need comparable strength. In terms of permeability (or permeance), finishes should permit movement of vapor out of the wall. Since moisture flows from warm to cool, the finish on the warmer side of the wall would ideally be *semi-permeable*, while the finish on the cooler side of the wall would be *permeable* (see the moisture guidelines on page 60).

Different plasters and stuccos fulfill their crucial functions quite differently. Mixes containing cement are the strongest, and are a good match for load-bearing walls because they resist the loads imposed on them. Yet remember that straw bale wall finishes need to be vapor-permeable, allowing the movement of vapor out of the wall. Stuccos composed of only cement and sand are relatively vapor-impermeable. The addition of lime (in a ratio of 1 part cement, 1 part lime, and 6 parts sand) greatly improves vapor permeance while maintaining good compressive strength and resistance to weather. Lime and sand by themselves are not as strong, but with slightly thicker layers, lime-sand plasters will meet most structural requirements, are very vapor-permeable, and offer good weather protection.

Even though earthen plasters have low compressive strength, when applied in layers of 3 inches or thicker, earthen plasters will probably achieve adequate structural strength. Their vapor permeance is higher than cement stuccos, but they are also the most susceptible to softening and erosion when exposed to rain, and therefore require the most maintenance or protection in the form of porches, intermediate roofs, and overhangs. Earthen plasters can be improved by adding larger quantities of straw, reducing the proportion of sand, and including some type of stabilization or cement. Thin coats of lime and sand can be applied over earthen plasters to increase their weather resistance.

STRAW BALE STUDIO

KNOWN TO SOME as the "Les Trois Trollettes de Troit +1," a group of four women that included owner Fran Lee and friends Deanne Bednar, Carolyn Koch, and Gregory Matthews, who joined together to build this 600-square-foot, timber-frame, straw bale building in Oxford, Michigan. Although none of these women is a professional builder, in combining efforts they could do much of the work themselves, primarily on weekends.

They sought guidance and help as needed, attending workshops around the country, including two weeks with Dutch master thatcher Flemming Abrahamsson. They enlisted a stone mason to teach them how to lay the foundation and another professional to help build the wood frame. With lots of volunteers, they stacked the bales, applied earthen plasters, and thatched the roof with water reeds they harvested locally. Cutting and gathering the reeds was extremely labor-intensive, but once the building was thatched, the thought of buying asphalt shingles seemed absurd. During the entire construction process, the group welcomed and taught visitors, school classes, and curious volunteers of all ages. Eventually Detroit Edison donated photovoltaics. As Fran Lee described the building, "This is supposed to be like a warm hug when you walk in."

Facing page: Exterior. Manufactured materials were avoided whenever possible, by relying on local and natural materials when they were available. Rather than "construction," the experience became a process of sculpting a building that meshes with nature.

Below: Roof overhang. The earthen plaster on the exterior walls required no additional stabilization thanks to the generous roof overhangs and a high foundation stem wall, which allow little or no precipitation to hit the wall.

THE SAVE THE CHILDREN OFFICE BUILDING

Emiliano Lopez.

Ciudad Obregon and its surrounding area in Sonora, Mexico, has been a second home for us since we began working there in 1995. Originally we were invited by the organization Save the Children to help develop a straw bale housing program for poorer communities in the area. Eventually it became obvious that we needed a demonstration building that would show the incredible potential and beauty of the natural materials we wanted to use. Even though straw, clay, and a local reed called "carrizo" are abundant, cost little or nothing, and are easy to work with, for a long time they have not been accepted as "real" materials like cement or brick. We needed a building people could visit and leave saying, "Wow, I want one."

This 5,000-square-foot office building was a joint effort between Save the Children, ourselves, and a team of mostly young, unskilled women and men, trained on-site by two brothers who are seasoned builders, Emiliano and Teodoro Lopez. Only the basic floor plan was established at the outset, then a steady flow of experimental ideas took over. Everyone became part of the creative process, and essentially the building was designed as it was built.

Most of the work was done by hand, using simple tools. The main power tools were a hammer mill to chop straw and a ½-inch drill for mixing finish plasters. The building's overall cost was approximately half that of conventional brick and concrete construction.

For us, this project exemplifies a different way of building—where people and relationships are a vital part of the process. Through the friendship of working together we found a magic that enriched and empowered us all.

Carrizo reeds form the arch of the entrance. The blue is a traditional Mexican shade known as *azul anil.*

Facing page, and above: To facilitate passive cooling we utilized a traditional Mexican courtyard design. The inner courtyard is surrounded by a palm-thatched porch or "portal." We used straw bales for the exterior walls, and straw-clay blocks for the courtyard and interior walls. A thick substrate of straw-clay plaster filled out the walls. We then finished the exterior with a lime plaster, subsequently painted with local pigments and colored clays using "fresco" techniques.

Overleaf: Reception room. Most of the interior walls were finished with a combination of clay/wheat-paste paints and frescoed plasters. The vaults in several rooms were formed from multiple layers of carrizo reed with a six-inch layer of straw-clay above, for insulation. The concrete floors were scored, then acid-stained to look like tile and stone.

Above right:
Teodoro Lopez.

RECIPE FOR CLAY PAINT OR "ALIS"

This method is about as easy as it gets for applying thin layers of clay to plastered walls, drywall, or other surfaces. The first time that we extensively used this formula was on the interior of the Save the Children office building, where we used a combination of local clays and pigments with splendid results. It creates a wall surface that is much more pleasing than the rubberized surface of many modern paints.

Step 1: To make a starch paste.

Add 1 part white flour to 2 parts cold water, and set aside.

Boil 1½ parts water. When boiling, add the above mixture of flour and cold water.

Turn heat to low, continue to cook until the mixture thickens. Continually stir the bottom to prevent burning.

Step 2: To make the paint.

For every 1 part starch paste dilute with 2 parts water. That will make the paste liquid enough to permit adding the other ingredients.

Next add enough colored clay, or white kaolin clay with pigments, to achieve a consistency that will spread easily with a brush. It is difficult to describe, but we look for a mixture that will cover in two coats. The right consistency will be like thick cream. Mica can be added to simulate clays that have naturally occurring mica. Fine (screened) chopped straw can also be added. If adding either or both, add less clay because the mica and straw too will thicken the mix.

Step 3: Application and polishing.

Apply with a brush, and when the paint has set but is still moist, use a damp (not wet) tile sponge to polish the surface, removing excess dust and revealing the straw and mica. A plastic lid with the edges cut off can also be used as a flexible scraper to further polish the surface.

CASAS QUE CANTAN:
A Women's Building Project in Mexico

On the outskirts of Ciudad Obregon, in a poor rural community called Xochitl, a group of women joined together to help build each other's homes. Most people in Xochitl live in what are called *Casas de Carton*—shacks made from corrugated black-asphalt panels and other scraps. In contrast, the women chose for their project the name *Casas que Cantan,* which translates as "Houses that Sing." Their one-room homes, approximately 300 square feet in interior area, have cost about 500 U.S. dollars to build; most of the funds for construction of these houses have come from donations made to our nonprofit organization, The Canelo Project, by straw bale builders in other parts of the world, including the United States, Canada, Europe, and Australia. In continuation of our work with Save the Children, we have helped organize the project and provide some of the training for the women.

Right: Rebecca Lopez de Vargas, one of the driving forces behind the women's group, tying carrizo reed to the exterior the straw bale walls. The reed is used instead of rebar pins in the middle of the bales.

Above: Juanita Morales de Lopez and Athena Steen applying a staw-clay plaster that will be finished with a lime plaster.

Above: Applying a finish coat of lime plaster over a base of straw-clay plaster.

Left: The window opening was shaped and molded with the base coat of straw/clay plaster. It is finished with lime plaster that was frescoed with traditional Mexican blue pigments.

MYRA VALDEZ'S NEW HOME

There was a shack pieced together from old metal, cardboard, black asphalt sheets, and other discarded scraps of material, typical of much of the housing in the area. We thought of going in, but knew the oppressive heat would be even more unbearable inside. Next door was the building we had come to see, one of the first *Casas que Cantan* houses built by the women. With its soft-colored, lime-washed walls and hand-shaped door and window moldings, the new house stood in striking contrast to the shack. A curving band of river stones added protection and texture at the base of the earthen- and lime-plastered walls.

The door opened and out stepped Myra, the owner. Smiling, she shyly welcomed us all in. Amazingly, it was cooler inside than outside, and everyone sighed with relief as we escaped the heat into Myra's cozy room. A flagstone-patterned floor made of recycled pieces of concrete was refreshingly cool underfoot. The carrizo-reed ceiling cast a golden glow on the plastered straw bale walls. Simply decorated, the two beds and other basic furnishings had obviously been positioned with great care.

"Oh, how wonderful!" exclaimed a visitor from the States. "What do you think of it?" Myra lowered her head momentarily, then with her large, dark eyes she reflectively scanned her new home and shyly replied, "You know, I always thought to have something this beautiful you had to have a lot of money. But now I know you don't. You just have to be willing to work for it."

Above: Myra Valdez mixing straw-clay plaster by hand.

Top: The door and window openings were thickened with plaster molding to give more depth to the otherwise plain exterior. Tinted limewash lechada over the plaster gives the soft yellow color.

Above: The interior walls are also colored with a tinted limewash. The floor is made from recycled concrete.

ELIZABETH NUZUM'S STUDIO

Alamos, Sonora, is an old silver mining town in the foothills of Mexico's Sierra Madre Mountains, famous for its many restored colonial mansions. Elizabeth, who also owns one of the most beautiful restored adobe buildings in town, designed and supervised the construction of this 1,000-square-foot thatched-roof, straw bale studio on the outskirts of town. We worked on the early phases of the building, and it was the first time that we used an exterior pinning system, with trunks of a small local tree called *barra blanca.* In addition to palm thatch, many local and traditional materials were utilized, including several different colored clays for the interior plastering and floor. All the woodwork and ironwork were done by local craftspeople.

This thatched roof, made from locally gathered palm, integrates the traditional roofing style of the area with thick-walled construction. Shutters and hardware were made by local Alamos craftspeople.

RINA SWENTZELL

THE SWENTZELLS' GUEST HOUSE

THIS PASSIVE-SOLAR, 350-square-foot cottage was built by an acupuncture student from Germany named Jorge who was unable to find an affordable place to live in the Santa Fe area. He offered to build a small house on Athena's parents land in exchange for living there during his three remaining years of school. He built the basic structure and later Ralph and Rina Swentzell added porches and took care of all the finishing details. Both the interior and exterior walls are finished with earthen plaster. A casein and linseed oil emulsion was used to stabilize the exterior plaster. It is a remarkably compact and efficient little building. The ground level has an earthen floor, with a living area, kitchen, and a small shower, with the loft above for sleeping.

Right: The compact space is amplified by a loft. Earthen plasters and floor complete the interior.

Facing page: Exterior, shown after the completion of the plastering session.

NÖEL BENNETT'S HOUSE

Facing page: Combined living, dining, and kitchen area.

Located at 8,000 feet in elevation on a beautiful and remote site in the Jemez mountains of northern New Mexico, this passive solar, infill straw bale house was built by Nöel Bennett and her late husband Jim Wakeman, with design help from architect Michael McGuire. They wanted to build a house that would not dominate its surroundings and that maintained a low profile. Their intent was to blur the line between humans and nature, designing the house to utilize seasonal energies for natural heating and cooling. Before building, they spent a year on the site to study the cyclical angles of the sun and the other subtleties. The effectiveness of their planning was tested the first winter when the heating system failed and outside temperatures stayed below zero for over a week. The house remained comfortable.

Exposed Parallam trusses combine with corrugated metal ceilings in a way that encourages one to explore how different materials can be combined. Generous overhangs are provided by the metal roof, affording good weather protection for the straw bale walls. The changing colors of the sky and variations in the surrounding meadow are reflected off the ceiling into the interior of the house. Another unexpected benefit of the metal ceiling is that it reflects rising heat back to the space below.

Russ Betts, an old friend of Nöel's and her current partner, is completing the finishing details of the home.

MEXICAN COUNTRY STYLE

A pitched metal roof blends visual continuity with traditional high mountain architecture of New Mexico, and provides several practical benefits, allowing for generous overhangs that protect the walls from rain while encouraging entry of sunlight in winter and keeping it out in summer.

Above: Nöel Bennett and Russ Betts.

Right: The kitchen.

Facing page: This south-facing passageway connects the two wings, and provides much of the solar gain needed to heat the house.

MOISTURE GUIDELINES FOR STRAW BALE HOUSES

One of the major concerns with straw bale walls is moisture. Straw, like wood or adobe, will deteriorate when exposed over time to excessive moisture. Straw is a magnificent building material that can provide increased levels of comfort and aesthetic appeal, but it must be kept dry. Likewise, good indoor air quality depends on healthy walls. Fungal growth and mold can result in severe health problems.

1. Avoid building moisture into the walls

Build only with good-quality bales that have always been kept dry, both before and during construction. Bales that have gotten wet, risen above an acceptable moisture content, and then later dried out should not be used. Straw fibers easily deteriorate during the drying-out process. If possible, build the structure and roof first.

2. Reduce exposure to rain

Driving rain is responsible for many of the moisture problems that occur in all types of buildings. Keeping water off the walls will greatly reduce potential problems.

- Select a site that is sheltered by natural features and not subject to strong winds and driving rains.
- If possible, plant trees and shrubs for windbreaks, and incorporate other types of screens.
- Good site drainage around the perimeter of a building is needed to direct away surface runoff.
- Roofs need generous overhangs. Wrap-around porches are even better than eaves for rain protection.
- The top of the foundation should be a minimum of 8 inches above grade, and exterior stucco or plaster should also stop well above grade. Protection from rain splash-back at the base of walls is essential.
- Gutters with downspouts can prevent roof runoff from falling near the base of walls and then blowing back against the walls.
- Use water-repellent coatings that are also vapor permeable on exterior stucco and plaster to prevent absorption of rainwater. Siloxane is one product that meets these criteria.
- Provide good window detailing, using sloped sills; head, jamb, and sill flashing; and drip edges to shed water away from the wall.
- Flash and/or caulk all joints between different materials to prevent unwanted intrusion of moisture or critters.
- Use berms to reduce the impact of snow, rain, and wind.

3. Avoid the use of vapor barriers

Vapor flows from warm to cold, and walls need the ability to dry. Since drying predominantly occurs through vapor diffusion, an impermeable barrier or coating can be detrimental. The vapor permeance or perm rating of both interior and exterior wall finishes needs to be chosen with due regard for each specific climate. Otherwise, vapor can be trapped within the wall, where it will condense. A general recommendation would be to make the finish on the warmer side of the wall *semi-permeable,* so that it slows the flow of vapor in the wall, while the finish on the cooler side of the wall should be *permeable* so that vapor can flow through.

Vapor permeability is typically rated in "perms."

Impermeable materials = 1 perm and less
cement stucco (1 part cement, 3 parts sand) = 0.5 to 1.0 perms

Semi-permeable materials = 1 to 6 perms
cement-lime stucco (1 part cement, 1 part lime, 6 parts sand) = 4 to 6 perms

Permeable materials = 6 to over 50 perms
lime plaster (1 part lime, 3 parts sand) = 8 to 12 perms

4. Provide a "continuous" air barrier to block leaks

Air flow or leakage occurs through openings (holes, cracks, etc.), and can carry 100 times more moisture than what might be generated by vapor diffusion through the wide expanse of wall and ceiling surfaces. Improperly sealed joints at penetrations in the ceiling, floor, and wall junctions as well as cracks around windows, doors, and electrical boxes are common locations where these problems occur. A continuous air barrier strategy relies on good detailing and a variety of materials to be effective. The stucco and plaster skins typically used on straw bales play an important part in establishing this continuous barrier.

5. Protect the bales from moisture at their connection with the foundation

Keep the bales above floor level, with raised sills or toed-up foundations, to help prevent damage from plumbing leaks and minor floods. The tops of foundations also need to be moisture-proofed: install a capillary break between the foundation and wall structure to prevent wicking of moisture up into the bales.

6. Monitor the walls for moisture

You can make your own moisture meters following instructions in issue no. 22 of *The Last Straw Journal* (see page 114). Moisture can be monitored either by checking moisture content or relative humidity. Straw that is under 15 percent in moisture content and lower than 75 percent in relative humidity can be considered sound and unaffected by moisture.

STEVE AND NENA MACDONALD'S HOUSE

Above: This simple exterior form is dictated by the shape of the shed roof. The shade structure is made from local branches.

Facing page: Interior. The trunk of an alligator juniper serves as a coat rack. The floors are soil-cement. Steve designed the interior walls as movable partitions.

ONCE WHEN we were talking with a contractor about what factual information he would like to see changed when we did a revision of *The Straw Bale House*, he told us, "Get rid of that $7-per-square-foot nonsense." Well, fact of the matter is that natural-building pioneers Steve and Nena MacDonald of Gila, New Mexico, built their post-and-beam straw bale home in 1988 for $7.50 per square foot. This small and simple home, perhaps not likely to turn up in *Architectural Digest*, is comfortable and meets their needs. They kept their costs low by organizing work parties with neighbors and friends, using mostly recycled or gathered materials and investing many hours of their own hard work. Their home exemplifies seven guidelines Steve listed in the book *Build It With Bales*, which he co-authored with Matts Myhrman:

Keep it small.
Keep it simple.
Build it yourself.
Stay out of debt.
Use local materials.
Be energy-conscious.
And: Make yourself a home, don't just build a house!

SUE MULLEN AND OSCAR DAVIS'S RESIDENCE

NEIGHBORS OF Steve and Nena MacDonald, Sue and Oscar have constructed several straw bale buildings on their site in southern New Mexico. This two-story structure serves different functions. The south-facing wing is a combination solarium and bathing area, while the other part of the bottom floor includes an office and shop. The upper floor is used mostly as a dance hall/ meeting room, with Oscar and Sue's bedroom on one end. They did much of the work, with help from two part-time carpenters, using mostly salvaged and recycled materials gathered over the years. Total cost of their house was about $25 per square foot.

Dance hall/meeting room. Unplastered walls make for great acoustics.

Above: The two-story dance hall exterior, south-facing. The lower portion is a solarium and bathing area.

Left: Interior of the solarium.

THE LAMA AND COLOMBO HOUSE

BUILT AS A spec house in Santa Fe, New Mexico, by Tony Perry and Ted Varney, this straw bale home has infill bale walls with a structure of concrete-block columns and wooden beams. The exterior finish is cement stucco, while the interior walls are finished with a unique blend of gypsum plasters and vermiculite. The house was designed to have radiant-floor heating, but during the first winter the system activated only twice, thanks to the combination of insulating straw in the walls and solar gain from the small south-facing sunroom.

Interior arch finished with blended gypsum plasters.

RECIPE FOR THE INTERIOR GYPSUM PLASTER

Combine:
1 part Red Top Dual Purpose Gypsum Plaster
2 parts vermiculite
3 parts Gypsolite or Structolite

Part of the interior volume of space is used efficiently to create a loft over the bedroom.

GINNA SLOAN'S HOUSE

Designed by Jan Wisinewski, this infill straw bale home was built by Ginna Sloan of Santa Fe with the help of local builder Stefan Bell. Both the interior and exterior plasters were prepared with local clays, as was the earthen floor. The majority of straw bale homes built today rely so extensively on manufactured materials that sometimes the bale walls seem incidental. For convenience and cost, interior walls are commonly stud-construction covered with drywall, in stark and boring contrast to the subtle undulations of the plastered exterior bale walls. Joint compound and latex paint produce a very different effect. It is refreshing in Ginna's home to see the extensive use of adobe for the interior walls, not only adding beauty and texture, but also providing thermal mass to help regulate interior temperature swings. One of the ironies of our time is that adobe, made from soils with clay, has become one of the more expensive building materials largely because it is more labor intensive. We would offer the counter opinion that anyone can make adobes and learn to build with them almost anywhere in the world, any time, for essentially no cost.

Living area with earthen floor. The interior adobe wall along the left side adds texture while providing thermal mass. Functional as well as beautiful.

The main living area is warmed by the late afternoon light, which highlights the earthen floor and clay-mica wall finishes.

THE HOUSE BUILT BY BOB MUNK

THIS 3,800-SQUARE-FOOT house in Santa Fe, designed by architect Beverley Spears, is also reprised from our original book. We've included it again because it photographed so well and because of the beautiful use of earthen materials, particularly with the plasters. A fine finish coat of micaceous earth plaster gives subtle sparkle to the walls. According to very accurate records, the complete cost of the house was $110 per square foot in the mid-1990s. Bob Munk acted as his own general contractor. The house was constructed using straw bales as infill in combination with 16-inch core-filled concrete blocks for posts and an 8-to-10-inch-wide reinforced-concrete bond beam on top of the walls. Several of the interior walls are also made of straw bales. The entire house has adobe/earth floors. Bob eventually sold the house, realizing the kind of appreciation expected of any conventional building, proving that people needn't be so worried that they will be penalized later for an investment in a straw bale home.

Exterior. For the size of this building, it integrates well with its surroundings.

The entry porch was purposely kept low to create an intimate, intriguing space that serves both as a passage to the interior and a sitting area.

CAROL ANTHONY'S CLOISTER AND SANCTUARY

CAROL DESCRIBES her load-bearing straw bale cloister as an homage to the inner courtyards and atriums of eighteenth-century Italian villas, but we feel that what makes this place particularly special is the way it reflects so much of Carol herself and the kind of person she is.

The building is 400 square feet, and is surrounded by a straw bale wall incorporating a thatched, African-style adobe granary tower in one corner. The cloister was built by a group of friends in their spare time, with Santa Fe builder Ted Varney constructing the walls. In Carol's words, the process was informal and communal, " . . . built with friends, great humor, tequila, a few hugs, and some delicious ratatouille."

To honor her twin sister, who had died, Carol made a one-room addition to her main house as a sanctuary space that could be used for solitude and quiet. This was built by local builder Rob Thomas with our help. The sanctuary has earthen plasters on both interior and exterior walls. Woven reeds were attached to the existing cement-stuccoed wall as lath for the earth plaster.

Interior with truth window.

The inset door shows the depth of the straw bales.

PAULA BREYMEIER'S HOUSE

STRAW BALE BUILDER Tom Leucke, originally part of the Boulder group known as Strawcrafters, built this home for himself in Crestone, Colorado, before selling it to Paula. It is a hybrid, combining load-bearing and infill straw bale walls. The spacious volume within the steep roof structure contains a full upper story. The dormers add usable space and allow more light into the upper rooms. The exterior earthen plaster is stabilized with Portland cement.

Facing page: The 1½-story design with shed dormers on the south side allows for generous interior living space and sufficient solar gain to make the house comfortable.

Right: Interior stairs with sculpted platform. Stairs are detailed with local woods.

TOM LEUCKE

TOM LEUCKE

ROOFS FOR STRAW BALE HOUSES

The roof of a building is the most directly exposed to the weather, and must meet the demands made upon it. If rainfall is high, the roof needs to shed water away from the walls; where snow loads are heavy, roof framing needs to support that additional weight. When the sun is intense, the roof should provide shade and keep the heat from radiating against the walls. Heat gain or loss is greater through a roof than through the walls, so whether in cold or hot places, roofs need insulation that will keep heat out or in, as needed.

A roof is also very site-specific in terms of profile and presence, needing to fit the surrounding landscape in a way that is visually interesting without becoming the dominant feature. Some sites demand a low building while others can more easily accept one that is tall.

A roof's shape will in many ways be dictated by the living space it is intended to shelter. For simplicity's sake and ease of construction, the most practical choices are the gable roof or the shed roof (in drier climates). The hip roof is also an option in some situations. All are within reach of the owner/builder. For roof support, trusses are easy to buy or build in a variety of configurations, and they make very efficient use of material. For example, scissors trusses allow for a sloped interior ceiling as well as a sloped roof.

In general, roofs should slope at a minimum of 4:12—preferably 6:12, to improve rain control and reduce snow loads. In drier climates the slope can be as flat as 2:12, with a simple shed roof.

Gable roofs can enclose a large volume of space. One of the easiest ways to gain space in smaller buildings without enlarging the footprint of the building is to raise a gable roof slightly to create a half-story or a loft. Dormers are a little more complex to build, but they amplify the interior space and add significant amounts of light. Shed dormers are easiest to build.

With straw bale buildings, one of the roof's main tasks is to keep as much water as possible away from the walls and base of the building. Generous overhangs are essential in wet regions, and in very dry regions with little rainfall, they provide shading from sun in summer, yet allow winter sun to enter. When walls are not too tall, keeping the eaves low will make the overhangs even more effective. For instance, where we live, a 32-inch overhang on walls approximately 7½ feet high is not overly exposed to high winds and keeps the wall surface dry most of the time. Gable-end walls should have wider overhangs than eave walls, because the gables are higher. Gutters should carry water away from the building and keep roof runoff from splashing back against the base of the walls. In colder climates, due to the weight of snow and the extra stresses of freeze-and-thaw conditions, gutters need appropriate detailing.

Partial or wrap-around porches make great sense in many situations. A small porch, 6 feet in width, easily protects walls on the windward side in all but the most severe, driving rains. In addition to improved rain control, porches can increase the amount of useful space dramatically at a much lower cost than fully enclosed additions, providing screened living and sleeping areas, utility space, outdoor kitchens, storage, space for washers and dryers, and when combined with clear glazing, a greenhouse or solarium. They need not reduce light in the building's interior. Clear panels in the porch roof can replace solid ones; tempered glass panels from salvaged patio doors might be used. Where horizontal driving rain is common, enclosed porches with operable windows or sliding wall panels are appealing.

Least suitable for straw bale houses is unquestionably the "Santa Fe"-style of parapet roof, where the walls extend up past the roof to present flat surfaces directly to the weather. Such roofs offer none of the advantages we mention above, and are suitable only in the driest desert regions. Parapet roofs are functionally nonsensical—they merely imitate the style of another time and place. Parapets have their origins in masonry building methods such as adobe. In northern New Mexico, the "pueblo-style" parapet roofs can be traced back to when there were no other roofing materials available and roofs had to be made flat and covered with dirt. Masonry materials also have a tendency to absorb and hold moisture, making them especially inappropriate companions for straw walls.

Today, when a variety of roofing materials is available, as well as different designs to address local climates and conditions, it is best to have the tops of walls completely covered by the roof, with further protection provided by generous overhangs.

LAURIE ROBERTS AND LANE MCCLELLAND'S STUDIO

THIS COZY STUDIO in San Diego County creatively draws upon Lane's woodworking skills and Laurie's talent for metalworking. The studio has load-bearing walls and a central pole in the middle of the building to support the plywood decking, which is covered with felt. Thatch laid around the edges of the roofline adds a decorative and organic touch. The cement stucco on the outside of the building is richly colored with ferric nitrate. The studio is 450 square feet, and was built for a total of $3,400, or $7.50 per square foot. Since Lane and Laurie did all the work themselves—except for one wall-raising workshop, which generated net proceeds of $2,000—the studio cost the owners $1,400 in out-of-pocket expenses.

Facing page: The ceiling is made with willow branches with their leaves left on, sprayed with linseed oil as sealant. The floor is soil cement, colored with a diluted wash of ferric nitrate.

Below: Door and willow furniture made by Lane. An accidental addition of an extra partial course of bales during the wall raising created the fortuitous rise over the door. Below right: Laurie and Lane.

Straw Bale Vault and Cob Office: PERMACULTURE INSTITUTE OF NORTHERN CALIFORNIA

THIS COLORFUL vault in Point Reyes Station is a true vaulted structure, supported by the form of the arch itself. The bales are set between timber-bamboo ribs, paired internally and externally. Corrugated metal roofing protects the bales from weather. Over windows, transparent corrugated roofing allows light in. Bob Theis and Dan Smith designed and led the workshop that built the vault. The nearby office building, which was begun during a Cob Cottage Company workshop, has a back wall of light clay-straw and cob.

Bob, Dan, and Kelly Lerner have been pursuing small vault structures such as this for a number of years. Vaults are not easy to build, and require caution, as a collapse could have fatal consequences. In addition to structural challenges, there is the problem of rainproofing the bales. In this case, corrugated metal was a good solution. Cement stucco and other plaster finishes would most likely not provide the necessary protection, unless they were flawlessly applied and meticulously detailed. Aside from their vulnerability to moisture, straw bale vaults can be exceptionally strong and resistant to seismic forces. David Mar, a Bay Area structural engineer, devised a vault design using bales as compression struts. The bales are wrapped in wire mesh with lengths of rebar running outside as an exoskeleton. Tests conducted on this design showed it to be more than sufficiently strong to meet the minimum building code requirement.

Right: Penny Livingston and James Stark.

Top: Straw bale structural vault with corrugated metal roof.
Above: Interior of vault with exposed bamboo pins.

The office at the Permaculture Institute is a cob cottage with a flowing curved metal roof. The sculpted dragon on the right is a cob baking oven.

We really like cob as a material and the possibilities it offers in combination with straw bales. Cob is a mix of clay, soil, and sand blended with water and straw. The mixture is made stiff enough that it can be formed by hand into walls for buildings, seating, fireplaces, ovens, and moldings. We don't seem to be able to build a straw bale building without using cob in some capacity. It is labor-intensive, but can be a perfect material for many of the interior elements of a straw bale house, including walls and partitions.

JOSEF KASPAROWITZ

Above: The many curved walls in Ken and Polly's home are based on fractal geometry, which emulates nature through the repetition of similar shapes that lack straight lines.

KEN HAGGARD AND POLLY COOPER'S RESIDENCE, GUEST HOUSE, AND ARCHITECTURAL STUDIO

In 1994, wildfires in the San Luis Obispo area of central California destroyed the entire complex of buildings owned by architects Ken Haggard and Polly Cooper. Fire left a large inventory of dead trees available for use in construction of a new group of three different buildings that would serve as their residence, studio/office, and guest house. Turko Semmes built the guest house, and Richard Beller built the office and residence.

The studio/office has a heavy timber structure of Sargent cypress and Douglas fir. Ken and Polly used heavy timbers because it was more economical to mill big pieces of wood than small ones. In their design of the studio/office, they placed great emphasis on natural lighting. South-facing glass, unvented 12-inch-thick trombe walls, and 9-inch-thick water tanks provide winter heat.

JOSEF KASPAROWITZ

Skylights with "sky-lids" provide additional solar gain. The building is cooled by night ventilation and well-distributed thermal mass.

In the two-story residence, Ken and Polly decided to place the timber frame six inches into the interior of the straw bale walls, facilitating construction and exposing the beauty of the wood. Locating the frame on the inside also eliminated the need to carefully detail the interface between wood frame and straw bales, which can be difficult and a source of air leakage into a wall. The timber frame allowed Polly and Ken to incorporate a two-story curving bale wall without interruption on the north side. Eight-inch concrete blocks were used for shear walls, thermal mass, and decorative gates.

The two buildings, studio/office and residence, together total 3,200 square feet.

The 500-square-foot passive solar guest house was built from fire-damaged telephone poles and a truss-joist frame. In contrast to their residence where the timber frame is exposed to the interior, here the poles were placed outside the bales and then covered with stucco. Rice-straw bales in the walls were set on edge and also used for insulation in the roof, between manufactured TJI rafters.

All the new buildings are off-the-grid and designed for passive solar heating and cooling. Ken and Polly's complex provides good examples of the use of sustainable materials, photovoltaics, natural lighting, passive conditioning, and sculptural forms.

JOSEF KASPAROWITZ

Above: The interior of the guest house with a south-facing, 4-by-8-foot skylight with a sky-lid (moveable insulation) unit.

Facing page: The timber-frame structure is placed 6 inches to the interior of all walls.

NATURAL CONDITIONING AND THERMAL MASS

Natural conditioning refers to passively heating, cooling, lighting, and ventilating buildings without the use of mechanical systems or imported energy, instead relying on sunlight and fresh-air ventilation. If designed well, naturally conditioned buildings are inherently more comfortable and healthier than those that are mechanically conditioned. Passive solar and natural cooling strategies also maximize the effectiveness of other building materials utilized.

Viable passive strategies depend on providing good insulation toward the outside of the building and enough interior thermal mass to keep temperature swings within a comfortable range. Materials with good thermal mass are those that retain heat or coolness. Examples include masonry materials such as brick, concrete, stucco, adobe, earth, and the stucco skin on straw bale walls. Water also provides excellent thermal mass, with roughly two times by volume and four times by weight more thermal storage capacity than masonry materials.

Constructed thermal mass falls into two categories: *concentrated mass* and *distributed mass*. Concentrated mass is heavy and consolidated, located in specific areas of the building, and is usually more efficient at heating than cooling. Distributed mass is spread out over larger surfaces and is particularly good for night cooling, because that larger surface area facilitates gradual transfer of coolness from the night air, which offsets the heat of the day. The optimal thickness in this case is approximately 2 inches. Distributed mass has a relatively high capacity for absorbing heat as well. In a good design, these two forms of mass are used together.

Successful passive designs depend upon establishing the proportions of insulation and mass needed to maintain consistent comfort in a given climate and set of circumstances. Straw bale buildings by their very composition offer an excellent starting point, with their highly insulated walls and the distributed thermal mass provided by the stucco or plaster. With good solar orientation and some provisions for concentrated mass, they can perform beautifully. Holistic computer modeling can be very helpful in fine-tuning designs for maximum performance.

Because they know much more about natural conditioning than we do, this summary is condensed from an article written by Ken Haggard and Polly Cooper with Jennifer Rennick featured in the book, *Alternative Construction* (John Wiley, 2000). Ken and Polly's architectural firm San Luis Sustainability Group is located in San Luis Obispo, California. The house and office shown to the right is a very tangible example of passive principles at work in their designs.

Studio/office interior. Interior shear walls are built of concrete block and provide thermal mass needed for natural conditioning of the building.

JOSEF KASPAROWITZ

HEIDI SCHLECT AND STERLING KEENE'S GREENHOUSE COTTAGE

Above: Heidi, Shani, and Sterling.

Facing page: Heidi and Sterling's south-facing cottage remains warm and comfortable in the damp, cool climate of northern California. They have since applied a vapor-permeable, water-repellent coating to the exterior walls.

WE REALLY ADMIRE this cottage for its compactness, ingenuity of design, and finish details. At 20 by 28 feet (560 square feet), Heidi and Sterling consider the space sufficient for their needs. They wanted to do much of the work themselves, and with no prior experience in construction, they made the layout simple. Also, keeping the project small meant keeping expenses low, so they wouldn't need a mortgage and instead could spend only what they had saved. Despite the home's modest size, Heidi operates a part-time catering business from her kitchen.

This post-and-beam structure has bales sandwiched between the rafters on the side walls. The loft is comfortable for sleeping, with ventilation maintained by an operable skylight. The space under the loft includes closet, drawers, pantry, and bathroom with composting toilet. The interior is earth plastered, using soil excavated on-site during construction. No sand but only straw was added for the base coat, then the final finish was done by painting on a slip coat of *alis* made from colored clay from a local pottery store.

Facing page: This interior shows a rare combination of compactness, efficiency, and beauty.

Above: The sleeping loft, with webbing to keep the baby safe.

Right: Detail of drawers under the stairway.

JAN EDLSTEIN'S MEDITATION ROOM

THIS LITTLE building, built by John Swearingen of Skillful Means Builders, sits snugly into an urban backyard in Mill Valley. Jan, a therapist with the Jungian Institute, uses her round building as a meditation and yoga space, and for therapy sessions with clients. It is 13 feet in diameter, with a living roof. The room has an earthen floor, and the interior walls are plastered with lime.

FERROUS SULFATE COLOR FOR CEMENT STUCCO OR LIME PLASTERS

Ferrous sulfate will stain both cement- and lime-based finishes a beautiful mottled rust color. John Swearingen of Skillful Means Builders, who has applied more of this solution to straw bale walls than anyone we know of, uses the following procedure.

John applies the ferrous sulfate while the stucco is still green and damp, within days of having applied the final coat. The walls are dampened before application and he tries to keep them damp as he works. The powdered ferrous sulfate, which can be bought from agricultural outlets or nurseries, is mixed into a bucket of water until granules or crystals start falling out of solution (John tries to keep everything in suspension). Using a brush or roller to apply the mix, he makes sure that a section of wall is saturated before moving on. The wall will first turn green, and it will take a couple of hours before the rust color appears.

Usually John applies two to three coats, waiting a day between each. Adding white Portland cement or lime, or tinting the final surface, will give the final color a different cast. The color continues to deepen for weeks and months after it is applied. Excess crystals left on the surface will eventually be washed off or can be rinsed off.

One note of caution: Ferrous sulfate will leave a hazy film on glass; to eliminate that problem, John uses a product called Rain-X, which is sold in auto parts stores. Ferrous sulfate will also turn wood gray, which can be good or bad depending upon whether or not one wants a weathered-gray look for the wood.

DAVID WARKENTIN'S HOUSE

THIS 1,150-SQUARE-FOOT straw bale home in Sausalito incorporates many recycled and reused building materials. Redwood from old bridge timbers and wine-tank staves was used for trusses, purlins, and exterior siding. All exposed framing is Douglas fir from a reclaimed warehouse floor in Oakland. The ceilings are covered with Meadowood panel sheathing made from rye straw. The straw bale walls are not load-bearing: 14-inch I-joists were used in conjunction with a Microllam beam for the structure. A Microllam is manufactured from thin sheets of veneer structurally bonded together to make headers and beams capable of spanning much longer distances and supporting heavier loads. The building was wrapped with 16-gauge wire for seismic reinforcement, and covered with ferrous-sulfate-colored gunnite on the exterior and soil cement on the interior. The floor is also soil cement, with red clay troweled into the surface. All doors in the house are recycled, and the kitchen countertop and dining-nook table were made from pieces of salvaged bowling alley flooring. Windowsills and countertops are precast concrete. Because of the building's thick walls, bay windows in the main room enclose futons for use as sleeping areas, with built-in storage below. The house was designed by David Arkin (also principal architect for the Real Goods Solar Living Center in Hopland), and was built by Paul Aurell.

Facing page: Office in main house.

Above: Exterior entrance.

Left: Thick straw bale walls easily accommodate window seats or sleeping areas.

The 300-square-foot turtle-shaped guest house was also designed by David Arkin. Built by Bob McKinney in a remote portion of the Warkentin site, all materials had to be hand-carried to the location. Trusses fashioned on-site made a curved ceiling possible, and these were plastered to give their unique appearance.

THE REAL GOODS SOLAR LIVING CENTER

THE 5,000-SQUARE-FOOT Solar Living Center in Hopland represents a commercial-scale use of solar technologies and ecological building techniques. David Arkin was the project architect, working with Van der Ryn Architects, Jeff Oldham of Real Goods managed the project, and Bruce King oversaw the structural engineering. Built of rice-straw bales stacked within a structure of concrete columns and Glu-Lam beams, the walls were finished with a soil cement PISE finish that was applied by rammed-earth builder David Easton. All wood used in the building was sustainably harvested from trees growing within a 40-mile radius of Hopland, then milled and manufactured into beams.

CATHERINE WANEK

The overall curved design encourages passive solar heating and cooling, and the interior is mainly lit by natural daylight.

MICHELLE LANDSBERGER AND DEBRAE LOPES'S HOUSE

WHEN FIRST planning to build a straw bale house in Santa Cruz, Michelle and Debrae enlisted the help of their architect friend Kelly Lerner to conduct a workshop in order to build a simple structure so they could learn the basics. They built a one-room structure, which Debrae now uses as her jewelry studio. They continued by creating a design for a larger living space with help from Kelly and another friend, Blythe Campbell. The 768-square-foot building (24 by 32 feet) they are presently building has a 10-in-12 roof pitch to accommodate a loft with two rooms underneath. The small extension at the gable end is intended for additional kitchen space. A bath house will be built separately. The house will be heated by a radiant-floor system and a woodstove.

Construction is being done by Michelle (serving as contractor) and Debrae, with the help of friends. While visiting them we had our eyes opened relative to permitting costs in northern California. To begin construction of a "garage," they secured an initial permit that cost approximately $2,500. However, a permit that will allow the building to be completed as habitable will cost $10,000, with a time limit for completion—meaning they will need up-front money for both permit and construction costs before they begin.

Left to right: Debrae Lopes, Kelly Lerner, Michelle Landsberger.

Above: The exterior color was achieved by combining two soup cans of copper sulphate with two tablespoons of ferrous sulfate in 2½ gallons of water.

Right: Wall niche.

EAST BAY WALDORF SCHOOL

THIS 1,100-SQUARE-FOOT building in El Sobrante serves as a combination woodworking and gardening classroom. A project of Vital Systems, designed by Greg VanMechelen and Christopher Day and built by Tim Owen-Kennedy, this playful building exhibits the creativity possible with straw bale walls while providing the benefits of natural insulation and breathable walls. The basic structure is a timber frame, and the classroom has limewash earthen wall plasters and earthen floors, cob seating, and many recycled materials.

The lovely color of the exterior finish combines colored clay, lime, sand, and fly ash.

Irregular shapes of windows and doors.

The new straw bale cottage in the rear of a suburban Berkeley lot.

EUGENE DECHRISTOPHER AND NAN AYERS'S COTTAGE

BERKELEY IS A densely populated urban area with an increasing number of apartment buildings, to keep pace with the demand for housing. Eugene and Nan realized they really didn't need all the space in their house, so they decided to build a smaller straw bale cottage for themselves in the backyard and to rent the main house. The project was challenging from a design standpoint due to the small size of the yard, the close proximity of surrounding apartment buildings, and the required legal setbacks. Because of their width and bulk, straw bales are not typically viewed as a material to use in small spaces, but with care the cottage was shoehorned into a tight space. The footprint of the structure is 20 by 22 feet with a total livable space of 850 square feet, which includes the second story. The cottage interior feels very spacious, and the bale walls create an aura of privacy and psychic calm amid urban commotion.

HOW MUCH DOES A STRAW BALE HOUSE COST?

This is one of the questions that we are most frequently asked. We typically respond, "It depends." We have built 300-square-foot homes in Mexico for $2 per square foot, we know that Steve and Nena MacDonald (see page 62) built their 800-square-foot house for $7.50 per square foot, and we've seen houses that cost as much as $385 per square foot. Why such a range? The variables are many, including the size and complexity of the building, location, architectural fees, types of materials, and who does the work.

In our experience, of all factors size has the greatest impact on cost. Most people want the biggest house they can get for the cheapest price per square foot. However, using cost per square foot as a measure can be very misleading. A well-built small house will cost far less per square foot than a mediocre big one. Moreover, 6,000 square feet at $100 per square foot ($600,000) represents much more money than 900 square feet at $200 per square foot ($180,000).

Small buildings tend to have simpler designs that require fewer materials and take less time to build. Plans can usually be drawn up by an owner/builder and submitted directly to the building official, and with smaller projects it is possible for owner/builders to do more of the work themselves. Using recycled and local materials is also easier on a smaller scale. Most importantly, a small house can more often be completed without leaving the owners in an exasperated state, on the edge of divorce or in financial ruin.

With small houses, a cost effective and realistic approach can be to combine the efforts of owners and sub-contractors. For instance, Complete Owner Builder Systems of Santa Fe, New Mexico, participated in the construction of two identical 1,450-square-foot houses. The first was entirely contractor-built, and cost $98 per square foot. On the second house, the owner sub-contracted the layout, excavation, concrete, electrical, plumbing, and plasters, but did the rest of the work himself. At $53 per square foot, his house cost approximately half that of the same, fully contracted structure.

John milled these beams from a Monterey pine planted by his father the year John was born.

JOHN HAMMOND AND SAYAKO DAIKIRI'S STRAW BALE COTTAGE

HAVING BEEN PART of the early development of passive solar housing in northern California, architect John Hammond was frustrated by the almost universal use of standard construction materials, especially wood-intensive stud-wall framing. He wanted to work with simpler and more local materials that could also help reduce costs. While visiting China, John was inspired by the aesthetic possibilities of indigenous materials used with care. He returned to California's Sacramento Valley and, remembering having read the 1973 edition of the book *Shelter,* he decided to build an experimental, low-cost cottage using straw bales, which are widely available in the region. He built an 18-by-22-foot building with a simple post-and-beam framework using bales as infill. The cottage was described by Gary Strang in a 1983 issue of *Fine Homebuilding.*

Fourteen years later, after teaching in Japan, John came back to California with his new wife, Sayako Daikiri, and decided to expand the original cottage by adding a straw bale wing. This new section includes a kitchen and loft area, and is also used as a living space.

Above: A porch and passageway connect the new addition to the original studio in an L-shaped configuration.

Right: John's original building, as featured in *Fine Homebuilding* in 1983.

THE HERMITAGE AT SANTA SABINA RETREAT CENTER

THE SANTA SABINA Retreat Center in San Rafael, run by the Dominican nuns, regularly hosts contemplative retreats for groups. This beautifully detailed cottage was built to provide an additional space for individual retreats. Designed by Dan Smith and Associates (DS&A), the cottage was the focus of a week-long workshop featuring David Eisenberg as well as John Swearingen of Skillful Means Builders, who completed the construction. Walls were built as load-bearing elements, with box beams set on top of the bales; then the walls were compressed, and 4-by-4 posts were inserted. That design takes advantage of the lateral stability and strength of load-bearing walls, and the additional posts prevent further settling in the walls over time, and provide convenient attachment points for doors and windows. Metal straps provide more seismic resistance.

Bob Theis, architect with DS&A, calls Santa Sabina an example of "jewel box" design and construction: "The smaller you keep a building the more you can polish it. People need to realize that it's not what you can afford, it's what you can inhabit. The rest becomes a burden."

Top: The window sill is made of tiles.

Left: Interior with sleeping loft.

FROG HOLLOW FARM

Facing page: "Farmer Al" Courchesne with a farmworker family.

Frog Hollow is an organic fruit farm run by Al Courchesne and Becky Smith, located in the delta region of San Francisco Bay in Brentwood. Frog Hollow accidentally became known for some of the best organic peaches in the world. When harvesting his first crops, Farmer Al, who didn't start farming until age thirty, didn't know that he was supposed to pick fruit green for storage and shipping, so he started picking his fruit ripe, only to end up with customers amazed at the sweetness, full flavor, and juiciness. Consequently, he's never changed his harvesting techniques.

Recognizing that the way he and Becky treat their workers is intimately connected with the quality of their crops, they wanted to improve the trailer accommodations typically used for Mexican farm workers. Having learned about straw bale construction, they contacted Dan Smith and Associates and worked out a plan for a small 24-by-26-foot house that included a loft space. Much of the prototype building was constructed during two weekend workshops, the first of which raised the walls and placed roof trusses. In the second, bales were custom-fit into the gable ends and between the trusses, where they supply critical heat insulation and serve as an acoustic barrier for noise from the highway that borders the farm. A scratch coat of stucco was also applied during the second weekend. The outside is stained with ferrous sulfate. The earth floor, made with a mixture based on psyllium seed, is durable as well as beautiful.

Above: Arched niche decorated with broken tiles in a mosaic pattern.

Above: A worker's house at Frog Hollow Farm. The cement stucco is stained with ferrous sulfate.

JOAN CROWLEY AND GENE SCHOTT'S HOUSE

A LICENSED CONTRACTOR and designer, Joan Crowley undertook the design and building of her house with the consulting support of Dan Smith and Associates. Joan and Gene needed a spacious home that could function as a gathering spot for their large combined family, which comes together frequently. Straw intrigued Joan with its aesthetic possibilities, as well as its potential to insulate them from the climatic extremes of their exposed site. Although this is a large house, the proportions and arrangement of interior spaces make it feel comfortable and warm.

The building's structure is a simple 4-by-4 post-and-beam system commonly used by DS&A on most of their buildings. Because of the tall walls, long vertical lengths of rebar were used on the outside of the bales rather than internally, as is commonly done. The rebar lengths are placed on opposite sides of the bale and then tied together, two pairs per bale. Each length of rebar was also tied to a short length of rebar extending vertically up from the stem wall. When used this way they are referred to as "exterior pins," although they really form an exoskeleton.

Joan was integrally involved in every facet of designing and building this house.

Above: The kitchen, with ample space to accommodate the many family gatherings that Gene and Joan host.

Facing page: Branches used to cover the entrance porch provide shading, texture, and filtered light.

KEEPING CURRENT WITH STRAW BALE BUILDING DEVELOPMENTS

For an ongoing useful e-mail discussion group:

CREST's Straw-bale Listserv
To subscribe, send a message with *subscribe strawbale* in the subject line to: majordomo@crest.org
Archives of past discussions:
http://solstice.crest.org/efficiency/strawbale-list-archive/index.html

For coverage of straw bale construction news and developments around the world:

The Last Straw Journal
HC 66, Box 119
Hillsboro, New Mexico 88042
(505) 895-5400
fax (505) 895-3326
thelaststraw@strawhomes.com
www.strawhomes.com
$28 per year in the U.S.

For straw bale building code information:

Development Center for Appropriate Technology (DCAT)
David Eisenberg
P.O. Box 27513
Tucson, AZ 85726-7513
(520) 624-6628
fax (520) 798-3701
info@dcat.net
www.dcat.net

For workshops on straw bale building, earthen floors, wall finishes, and projects in Mexico:

The Canelo Project
Athena and Bill Steen
HC 1, Box 324
Canelo / Elgin, AZ 85611
(520) 455-5548
fax (520) 455-9360
absteen@dakotacom.net
www.caneloproject.com

For wall raisings, customized workshops, design consultations, and public presentations:

Out On Bale, (un)Ltd.
Matts Myhrman and Judy Knox
2509 N. Campbell, #292
Tucson, AZ 85719

In addition, for an extensive listing of straw bale and natural building resources, including books, videos, and human resources, consult the annual Resources issue of The Last Straw Journal.